Mastering Building Construction: Tips for Site Supervisors and Owner Builders

Mastering Building Construction

Mastering Building Construction: Tips for Site Supervisors and Owner Builders

a handbook of how to navigate building construction processes and avoid possible errors.

by

Ojonimi Adegbe

PREFACE

Building construction is an ancient industrial activity. Since the quest for alternative shelters apart from caves, trees, and other natural forms shared with animals, humans have been involved in constructing various kinds of buildings. As knowledge increased, construction methods, materials, and the strength and stability of buildings evolved. While most of these advances in building have become commonplace today, new high-tech methods continue to emerge.

In addition to providing shelter, security, and privacy, key issues in present-day buildings include aesthetics, stability, convenience, accessibility, and amenities. The building process has thus become complex, with procedures and processes ensuring functionality from design to completion.

The site supervisor plays a crucial role in monitoring activities on-site to ensure that procedures and processes are followed for a smooth, successful, and desired project completion. As a site supervisor, understanding fundamental site procedures and processes is crucial. Whether you're a seasoned professional or a newcomer, staying informed can prevent errors and omissions. Even experienced supervisors occasionally encounter challenges, while the uninformed or inexperienced supervisor can easily fall prey of the wiles of contractors, subcontractors, and workmen.

While building processes in tropical regions share similarities, it's essential to recognize the nuances based on building type, size, and plan. In this guide, we'll outline the construction of a typical residential building, providing valuable insights applicable to any construction project.

This comprehensive guide outlines key activities, standard practices, processes, and potential pitfalls observable in the construction of buildings in particularly in the tropical region. It serves as a valuable reference for both novice and seasoned site supervisors, as well as other site workers. Additionally, the book functions as a practical checklist to prevent omissions and errors during construction. Written with owner-builders in mind, it provides essential insights for successful project management.

ACKNOWLEDGEMENT

I wish to express my gratitude to the management and staff of Synertech Properties Limited, whose inspiration and unwavering commitment during my tenure served as the catalyst for this book. To my beloved family members, especially my wife, Ndiana, I am deeply indebted for their unwavering love, which shaped my journey and enabled me to complete this work.

Lastly, I extend heartfelt appreciation to the divine source—God—for granting me life, intellect, and the energy to embark on and successfully finish this endeavor.

DEDICATION

This book is dedicated to my beloved wife, Ndiana. Her unwavering support and dreams have found fulfillment through my work.

CONTENTS

CHAPTER 1

PRE-CONSTRUCTION

Like any other technical project, building construction project can be divided into phases. Like any technical project, a building construction project is divided into several phases. One of the most critical phases is the planning or pre-construction phase. Before starting the actual construction, thorough pre-construction activities are essential. Key activities in this phase include reviewing building plan, obtaining building permits, awarding and executing contacts, project planning and task scheduling. These activities are critical steps to ensure a successful kickoff of the building project.

KEY ACTIVITIES AND STANDARD PRACTICES

Reviewing Building Plan

This step involves a thorough review of the architectural and engineering plans to identify any discrepancies or potential improvements. It also includes verifying with local or state government building officials to ensure that the design meets regional building criteria and regulations.

When it comes to designing a house in the tropics, several crucial principles should guide your building plan:

> Site Orientation: The orientation of your home on the site plays a significant role. Consider prevailing winds and sun exposure. For instance, in the Caribbean (north of the equator), prevailing winds come from the east, and the sun path primarily passes to the south. In northern Australia (south of the equator), winds come from the northwest or southeast, and the sun path passes to the north. Aim to orient your building to minimize sun exposure while capturing prevailing winds. Prioritize rooms needing ventilation toward these winds.

> Shading: You can prevent walls and surfaces from overheating by providing effective shading while designing

the house. This helps maintain comfortable interior temperatures without relying heavily on air conditioning.

Wall Materials and Construction: Choose materials suitable for hot and humid climates. Proper wall construction regulates temperature and humidity inside the house.

Roof Design: Consider concrete roofs (in-situ or precast) that minimize heat gain. In-situ concrete roofs are popular for residential buildings applications.

Exterior Spaces: If the space permits, incorporate outdoor areas that enhance functionality and connect with nature.

Remember, collaborating closely with your architect or design professional ensures these principles are integrated effectively into your building project.

Obtaining Building Permit

A building permit is a government-issued document that allows individuals or organizations to start construction projects. It ensures that the proposed construction is structurally sound, fire-resistant, and designed with proper safety measures. Acquiring a building permit is essential for construction in planned communities. This process involves submitting plans for review and obtaining approval from the local building or planning department. These authorities assess factors such as building setbacks and other regulations that impact property development and house placement. The requirements for obtaining building permit vary based on jurisdiction and local laws.

Creating a site plan

Creating a site plan involves accurately locating the new house on the building site. This process includes surveying property line dimensions, property easements, topography, and the placement of all utilities (including temporary structures like sheds or site offices). If the site is undeveloped, it's essential to identify the water supply (e.g., well) and on-site toilet location on the plan. Properly orienting your home on the site is crucial, considering factors such as prevailing winds and sun exposure. The site plan serves as a valuable construction document for the site supervisor

Executing Contracts

Most contracts with contractors and sub-contractors can be made fairly simple and straight forward. Whatever the case, there should be a written or verbal agreement and construction document which are required to fulfill the contract. Contract and construction documents are important references for the site supervisor. When it comes to executing contracts in construction, here are some key points to consider:

Simplicity and Clarity: Contracts with contractors and subcontractors should be straightforward and easy to understand. Clear language helps prevent misunderstandings and ensures everyone is on the same page.

Written or Verbal Agreements: While verbal agreements can be valid, it's always better to have written contracts. Written agreements provide a detailed record of terms, responsibilities, and expectations. They also protect both parties in case of disputes.

Construction Documents: Alongside the contract, construction documents play a crucial role in managing building projects. They provide detailed information for construction, ensuring everyone involved is on the same page. These documents include plans, specifications, and any other relevant details related to the project. They serve as references for the site supervisor during construction.

Creating Project Schedules

This is the process of listing out the tasks in a chronological sequence so that tasks can be carried out in their right order. Project schedules also highlight project timelines, milestones, and key dates. Scheduling affects total time of projects and often times scheduling mistakes can cause extra delay which may lead to extra development costs. It is the work of the site supervisor to detail and update the contractor/sub-contractors about site task schedules

STEP BY STEP PROCESS
The following is a typical step by step pre-construction process.

1. **Review Plans and Site Assessment:**
   a. Review the architectural plans to ensure the building can be properly sited on the property. Consult with the building official or relevant authorities.
   b. Make necessary adjustments or modifications based on feedback from a design professional.

2. **Site Planning and Design:**
   a. Create a detailed site plan that includes property boundaries, utilities, topography, and other relevant features.
   b. Identify the location of utilities (e.g. water supply, sewer lines) and any temporary structures (e.g. sheds, site offices).

3. **Obtain Building Permit:**
   a. Apply for a building permit from the local government agency.
   b. Ensure compliance with local laws, regulations, and construction ordinances.

4. **Finalize Detailed Drawings and Specifications:**
   a. Develop comprehensive construction drawings and specifications.

    b.   These documents guide the construction process and provide essential details for contractors.

5. **Task Schedule and Timeline:**
   a. Create a project schedule that outlines tasks, milestones, and deadlines.
   b. Consider factors like weather, availability of materials, and subcontractor schedules.

6. **Contractual Agreements:**
   a. Forge contracts with professionals (e.g., architects, engineers), contractors, and subcontractors.
   b. Specify roles, responsibilities, payment terms, and dispute resolution mechanisms.

7. **Project Commencement:**
   a. Set a clear start date for construction.
   b. Ensure all necessary preparations are in place before breaking ground.

POSSIBLE ERRORS

At the pre-construction level, several common mistakes can significantly impact the success of a project. Here are some key areas where errors often occur:

1. **Poor Planning**: Inadequate planning is one of the most common mistakes. This can lead to delays, cost overruns, and other issues down the line. Failure to account for site-specific factors such as topography, soil type, and utility locations can result in construction challenges. Underestimating costs or failing to allocate funds for unexpected expenses can lead to budget overruns. Proper planning should include detailed timelines, resource allocations, and budgets to avoid these pitfalls.

2. **Poor Documentation**: Keeping complete and accurate records of all project aspects is essential. Incomplete documentation can lead to misunderstandings and costly

errors. Utilizing document management systems can help streamline this process and keep all stakeholders informed.

3. **Omission of tasks in drawing out schedules**: Important tasks might be left out, leading to potential delays or oversight in project execution.

4. **Setting unrealistic timelines:** Underestimating the time required for certain tasks can result in scheduling conflicts and project delays. Overly tight schedules that don't account for potential setbacks can compromise resource availability and project completion. It's important to set realistic timelines and use project management tools to forecast potential delays.

5. **Errors in contracts:** Ambiguous or incorrect statements in contracts can lead to misunderstandings, disputes, or legal issues.

6. **Choosing the Wrong Subcontractors**: The performance of subcontractors can greatly affect project outcomes. Thorough vetting during the bidding process and clear legal agreements can help mitigate risks associated with subcontractor performance.

7. **Oversights in final drawings:** Missing details or errors in the final construction drawings can cause significant problems during the building phase, requiring costly and time-consuming corrections.

8. **Inaccurate site surveys:** Errors in site surveys can lead to issues with building placement and alignment, affecting the overall project.

9. **Failure to obtain necessary permits:** Not securing all required permits and approvals can halt construction and incur fines or penalties.

10. **Lack of communication with stakeholders:** Poor communication with contractors, subcontractors, and other stakeholders can result in misunderstandings and coordination issues. Effective communication is crucial. Misunderstandings or lack of information sharing among

team members, contractors, and stakeholders can cause delays and budget overruns. It's important to establish clear communication channels and ensure that everyone understands their roles and responsibilities.

11. **Safety Oversights**: Ensuring safety measures are in place is critical to avoid accidents and injuries. Regular safety inspections and adherence to safety protocols are necessary to maintain a safe working environment.

12. **Ignoring environmental impact assessments:** Overlooking the environmental impact of the project can lead to legal and regulatory complications.

13. **Inadequate risk management**: Failing to identify and plan for potential risks can result in project delays, cost overruns, and safety hazards. Not having a comprehensive risk mitigation plan can lead to unforeseen problems. It's important to conduct risk assessments early and regularly throughout the project, and to develop contingency plans to address potential issues.

By being aware of these common mistakes and taking steps to address them, project managers can improve the likelihood of a successful construction project.

TIPS AND OTHER CONSIDERATIONS

1. Written contracts are preferable to verbal agreements. A comprehensive contract should include the following specific items:
   a. Total cost of the contract.
   b. Cost breakdown by phase: If the work is divided into phases, provide a detailed cost for each phase.
   c. Cost breakdown of various components: Itemize the costs for labor, materials, transportation, and other aspects that contribute to the total price.
   d. Specifications of materials and equipment: Include details about the type of materials to be used, brands

and models of equipment or fixtures to be installed, and a general description of the work.

 e. Payment schedules: Clearly outline the payment terms and schedules.

 f. Costs for change orders: Specify how change orders will be handled and their associated costs.

 g. Responsibilities of each party: Define the general responsibilities of each party, such as site cleanup, safety precautions, and other relevant tasks.

 h. Project timeline: State the commencement and completion dates for the project.

2. Prompt notification to contractors and subcontractors who manage critical events should always be a priority.

3. The traffic flow of workers and materials on site should 1. Written contracts are preferable to verbal agreements. A comprehensive contract should include the following specific items:

4. Ensuring safety on site largely depends on regular clearing of rubble and adherence to other safety measures.

CHAPTER 2

CONSTRUCTION TO FOUNDATION FLOOR

The next phase of building construction involves constructing the foundation. In this chapter, we will outline the key activities, including site clearing, setting out, foundation excavation, and casting.

KEY ACTIVITIES AND STANDARD PRACTICES

Site clearing and Preparation

This step includes removing vegetation and sand filling where applicable. Additionally, it involves digging a well or borehole for water supply, providing an on-site toilet location, and erecting a site shed, store, or office.

Setting Out

Setting out involves marking the building foundation outline on the site using pegs and lines based on the construction drawings. This critical procedure must be supervised by a qualified engineer who can accurately interpret the drawings.

Foundation Excavation

A well-executed foundation starts with precise excavation. If the setting out is done correctly, the excavation will proceed smoothly without errors.

Foundation Casting

The type of foundation, which mainly depends on the soil type, should be specified in the construction documents. It is crucial to follow the details meticulously, as the building's stability relies significantly on the foundation's strength. Whether the foundation type is strip, raft, or pad, concrete and iron reinforcements are typically part of the details, which the workers and site supervisors should accurately interpret.

STEP BY STEP PROCESS

The following is a typical step by step process involved in construction to foundation level:

1. Construct an access road if necessary. Clear the land of vegetation, tree stumps, etc., and sand fill the land where applicable.
2. Dig a site well or borehole, establish a latrine location, build a site shed or office, and connect power and other necessary amenities to the site.
3. Supply necessary tools, equipment, and storable materials to the site.
4. Perform the setting out of the building foundation.
5. Excavate the foundation.
6. Apply blinding and anti-termite treatment where applicable.
7. Conduct ironwork (reinforcement) and carpentry work (formwork) where applicable.
8. Cast the foundation bed.
9. Build the foundation wall with blocks (if the foundation type is strip) and infill with concrete where applicable.
10. Fill and ram the hard core.
11. Screed and apply Damp-Proof Membrane (DPM) where applicable.
12. Conduct ironwork (reinforcement) and carpentry work (formwork) for the concrete floor.
13. Install plumbing and electrical piping where applicable.
14. Cast the concrete floor.

Adherence to these steps ensures a solid foundation, setting the stage for the rest of the construction process

POSSIBLE ERRORS

During the construction from foundation to floor slab level, several errors can occur. They include the following:

1. **Setting Out Errors:** Without proper supervision and crosschecking, setting out errors may occur. These errors can result in misaligned buildings concerning reference points and incorrect building perimeters.

2. **Concrete Mix Ratio Assumptions:** Incorrectly assuming the mix ratio for concrete can lead to problems. Specifications must be followed for different concrete mixes. The mix ratio specified for foundation beams might differ from that for the floor slab.

3. **Topography-Related Mistakes:** Construction errors can easily happen when dealing with different foundation levels, especially in undulating topography. If the site's topography isn't adequately considered (as noted in the construction drawings), it can lead to issues.

4. **Standard Measurement vs. Specified Measurement:** Assuming standard measurements instead of following specified measurements can be problematic. Always prioritize the specified measurements, especially if they include extra precautions. Standard measurements typically represent minimum requirements.

5. **DPM (Damp-Proof Membrane) and Anti-Termite Application:** Omitting or incorrectly applying DPM and anti-termite measures can compromise the durability and performance of the foundation and floor slab.

Attention to detail and adherence to best practices are crucial during this phase to avoid these errors.

TIPS AND OTHER CONSIDERATIONS

1. The site supervisor should ensure that the right quantity of materials needed for any stage of work is available (supplied) on site before work is commenced, this is to avoid shortage of materials, which will result to delay, and likely flawed concrete structures.

2. Quality of materials also can delay or lead to failure of structures. The site supervisor should be able to differentiate between sharp sand and filling sand, various types and sizes of iron rods and any other details of materials as specified in the construction documents.

3. Precision instruments like leveling instruments should be used for precision in measurements, instead of guesstimating or using simple manual means.

4. Proper means of obtaining the level of the concrete in casting the foundation or floor is essential to avoid waste of material and for a perfect leveled spread. Usually iron rods are cut to size of the depth required and attached vertically to reinforcement rods or form work around and within the perimeter of the area to be cast. The tip of the rods serves as level for the concrete poured during casting.

CHAPTER 3

CONSTRUCTION OF SUPER STRUCTURE

The foundation phase of building construction is critical, as any errors in dimensions or properties will impact the superstructure, which is the visible part of the building admired by people. This phase is where good workmanship becomes evident and tests the supervisory skills of the site supervisor. Key activities in this chapter include block laying, construction of lintels, columns, and arches, as well as plumbing, electrical conduit piping, and roof construction.

KEY ACTIVITIES AND STANDARD PRACTICES

Setting Out.

Setting out the superstructure is similar to the foundation stage but focuses on internal partitions. Precision instruments are often necessary, especially when dealing with non-rectangular rooms with acute angles and curves.

Block Setting

Block setting is a straightforward yet critical step in superstructure construction. Ensuring walls are plumb, erect, and straight requires skilled workmanship. Common materials include well-vibrated cement/sand blocks or burnt bricks. Nine-inch blocks are suitable for external walls, while six-inch blocks are appropriate for internal partitions with minimal roof load and no soundproofing requirements.

Lintels, Columns and Arcs

These elements are typically made of reinforced concrete or bricks. During this stage, iron benders and carpenters play crucial roles. Proper formwork is essential to avoid crooked columns, lintels, or arches. Careful concrete pouring ensures formwork stability and prevents material wastage during plastering.

Crooked columns/lintels/arches present difficulty and wastage of material during plastering.

Plumbing and Electrical Piping

Installing plumbing and electrical pipes within walls and under concrete floors requires precision. Detecting errors early—before casting the floor or plastering—is crucial to avoid material waste and time delays. Pipes are commonly made of high-quality PVC or galvanized steel.

Slab Construction

Most buildings, whether bungalows or multi-story structures, feature raised concrete slabs for water tanks or balconies. These slabs are supported by columns and beams cast simultaneously for structural uniformity. Accurate interpretation of drawings ensures proper slab positioning and dimensions. Cantilevered slabs rely on beams and columns for support.

Roof Carcass

Roof carcasses can be made of wood or steel. Wood is cost-effective and lightweight but must be treated to resist termites and rot. Steel rafters also require protective treatment against rust. Wall plates securely attach the roof to the building. While most roof carcasses are straightforward, complex designs may necessitate input from site engineers.

Roofing Cover

Aluminium sheets, available in long spans or step tile forms, are popular roofing choices. Other options include zinc, asbestos, ceramic tiles, concrete, and wood decking. After completing the roof carcass, precise measurements determine the required roofing materials. Ordering aluminium roofing sheets at this stage ensures error-free installation.

STEP BY STEP PROCESS

The following is a typical step by step process in the construction of the super structure:

1. Setting Out of Internal and External Partitions:
   a. Precisely mark the positions of internal and external walls based on architectural plans.
   b. Ensure accuracy for proper alignment during blockwork.

2. Blockwork to Lintel Level:
   a. Construct walls using well-vibrated cement/sand blocks or burnt bricks.
   b. Reach the height of lintels (horizontal beams) that support the load above openings (doors, windows).

3. Carpentry and Iron Bending:
   a. Prepare carpentry work for lintels, columns, slabs, and arches.
   b. Iron bending involves shaping reinforcement bars (rebars) for concrete structures.

4. Casting of Columns, Slabs, Lintels, and Arches:
   a. Pour concrete into formwork to create columns, horizontal slabs, lintels (above doors/windows), and arches.
   b. Proper formwork ensures accurate shapes and alignment.

5. Blockwork to Roof Level:
   a. Continue building walls to the desired height, typically reaching the roof level.
   b. Ensure plumb and straightness.

6. Plumbing/Piping:
   a. Lay plumbing pipes (usually PVC or galvanized steel) within walls and under concrete floors.
   b. Detect and address any errors before floor casting.

7. Electrical Piping:

   a. Install electrical conduits within walls and slabs.

   b. Coordinate with plumbing to avoid conflicts.

8. Frames and Sub-Frames for Windows and Doors:

   a. Construct frames and sub-frames to support windows and doors.

   b. Ensure proper alignment and fit.

9. Roof Carcass:

   a. Build the roof structure using wood or steel.

   b. Treat wood to prevent termite damage or rust for steel.

10. Roof Cover:

   a. Install roofing material (e.g., aluminum sheets, zinc, tiles) over the roof carcass.

   b. Ensure accurate measurements and proper installation.

POSSIBLE ERRORS

Errors that may occur at this stage include:

1. **Incorrect Partition Marking During Setting Out:** Errors can occur during the setting out of partitions, leading to incorrect room dimensions and layouts.

2. **Block Setting Errors:** Block setting errors are common, particularly related to the perpendicularity, levelness, and straightness of the walls.

3. **Errors from Plan Corrections:** Changes to the original plan that are not promptly communicated to the site engineer or supervisor can result in construction errors.

4. **Improper Embedding of Windows and Door Frames:** Windows and door frames may not be properly embedded or set in the correct positions, leading to alignment issues.

5. **Issues with Erectness of Columns and Straightness of Beams:** Columns may not be perfectly erect, and beams and lintels may not be straight. Additionally, the curvature of arches might be incorrect. These issues often stem from displaced formwork during casting or inaccuracies in formwork setup.

6. **Errors in Electrical and Plumbing Installations:** The positions, numbers, and levels of switches, sockets, and plumbing joints can be incorrect, leading to functional and aesthetic issues in the building.

TIPS AND OTHER CONSIDERATIONS

1. Any change in the house plans, especially affecting the external perimeter, necessitates an immediate redesign of the roof to fit the new layout. This helps avoid wasting time and resources.

2. Corrections to the plan drawings should be made after the first course of blocks is laid. At this stage, room sizes and partitions are visible, allowing clients to request adjustments without significant material waste.

3. Cement blocks are prone to breakage during transit, storage, and block setting. To minimize loss, avoid ordering too many blocks at once and reject wet blocks upon delivery due to their increased fragility.

4. Properly detached softwoods used for formwork can be reused for scaffolding or the same purpose, reducing project costs.

5. Mortar for bonding blocks typically requires a high cement-to-sharp sand ratio for strong bonding. To cut costs without compromising quality, clay can be added to the mixture, reducing cement usage.

6. The site supervisor should personally ensure that wood materials for the roof carcass are properly treated with preservatives to guarantee roof durability.

7. Mechanical and electrical (M&E) drawings should be detailed enough for plumbers and electricians to accurately lay pipes and position service junctions

CHAPTER 4

FINISHES, FIXTURES & FITTINGS

This chapter delves into the non-structural yet crucial aspects of a building, focusing on finishes, fixtures, and fittings. These elements play both functional and aesthetic roles in the overall design. Key activities discussed include plastering, ceiling work, flooring, wall finishes, and the installation of fixtures and fittings such as kitchen cabinets, wardrobes, doors, windows, plumbing, and electrical fitting.

KEY ACTIVITIES AND STANDARD PRACTICES

Rendering/Plastering

After the completion of the building carcass, the first finishing task is rendering, which uses plaster sand. The quality of plaster sand should be assessed for brightness and consistency. Effective rendering not only provides a smooth surface for subsequent painting but also addresses any errors in the block laying process.

Roof Ceiling

Ceiling options include tongue-and-groove (T&G) boards, plywood, plaster of Paris (POP), particle board, and asbestos. Ceiling work typically follows the rendering process and precedes the painting phase. If wall tiling extends to the ceiling, the tiling must be completed before the ceiling installation. The choice of ceiling material depends on the intended use—some are suited for internal environments, while others are designed for external applications.

Flooring

Floor coverings for concrete slabs can be wood laminate, tiles, terrazzo, marble, palladiana, or screed. Flooring is generally aligned with wall tiling. Palladiana, a decorative technique using broken tiles or marble, requires skilled craftsmanship. Skirting is an additional detail that enhances the floor's visual appeal.

Wall Finishes

Wall finishes include painting, wallpapering, tiling, plaster of Paris (POP), or paneling. Paint types include gloss, text-coat, and emulsion. Wall finishing is performed after the flooring is completed to ensure a polished and cohesive look.

Fittings and Fixtures

Fittings and fixtures encompass a range of components that contribute both functionally and aesthetically to a building. They include kitchen cabinets, wardrobes, shelves, burglary-proofing, doors, windows, plumbing elements, and electrical installations. The quality of these materials and the skill involved in their installation are crucial for the building's functionality and aesthetic appeal.

Kitchen Cabinets, Shelves, and Wardrobes: These are commonly constructed from various materials such as wood, particleboard, blockboard, or plywood, and may include various surface finishes like granite, wood, tiles, concrete, or Formica. Kitchen cabinets often feature countertops, which can also be made from granite, wood, or other materials. Wardrobes are similarly made from wood or hybrid materials, designed to provide storage solutions.

Doors: Doors serve multiple purposes including access, security, and aesthetics. They come in various types and sizes, such as panel doors, security doors, and flush doors. Each type has specific applications: panel doors are used for aesthetics and security, while flush doors are typically used for interior applications. Ensuring the proper manufacture and installation of doors is crucial for their effectiveness and durability.

Windows: The most common type of window is the aluminum-framed window, available in casement or sliding varieties. Other window types include louvered windows, glass-pane swing windows with iron or wooden frames. Aluminum windows are favored for their durability and ease of maintenance.

Burglary Proofing: These fittings are installed primarily for security purposes, commonly fitted to doors and windows to enhance the safety of the building.

Plumbing Fittings: Essential plumbing components include water closets (WCs), washbasins, shower trays, and bathtubs. These fittings are critical for the building's water supply and waste management systems.

Electrical Fittings: Electrical fixtures include wall and ceiling brackets, electrical sockets and switches, air extractors, intercoms, and water heaters. These components are integral to the building's electrical system, providing necessary utilities and functionality.

The site supervisor must ensure that all materials supplied are of high quality and that the installation of fittings and fixtures is carried out with precision and expertise to achieve the desired outcomes for both function and appearance.

STEP BY STEP PROCESS

The following is a typical step-by-step process at the finishes level:

1. **Plastering**:

    a. **Apply plaster:** Coat walls and surfaces with plaster to achieve a smooth and even finish.

    b. **Correct imperfections:** Address any flaws resulting from block laying to ensure a uniform surface.

2. **Ceiling Work**:

    a. **Install ceiling types:** Fit various ceiling materials such as tongue and groove, plywood, or plaster of Paris (POP), considering both aesthetics and functionality

    b. **Sequence of work**: Ceiling installation typically follows plastering and precedes painting.

3. **Flooring**:

a. **Choose floor coverings:** Select appropriate floor materials such as wood laminate, tiles, terrazzo, or marble.

b. **Ensure level installation:** Achieve a polished appearance by ensuring the flooring is level

c. **Align tiling:** Coordinate floor tiling with wall tiling where applicable.

4. **Wall Finishes:**

a. **Finish walls:** Apply wall finishes like painting, wallpaper, tiling, plaster of Paris, or paneling.

b. **Select paints:** Choose the right type of paint (e.g., gloss, emulsion) based on the desired effect.

c. **Sequence of work:** Wall finishes are applied after the flooring is complete to ensure a cohesive look.

5. **Fittings:**

a. **Install fixtures:** Fit kitchen cabinets, wardrobes, shelves, and other essential fixtures.

b. **Consider materials:** Use suitable materials such as wood, particleboard, or plywood, keeping aesthetics in mind.

c. **Include various fittings:** Install doors, windows, plumbing fixtures, and electrical fittings as part of this process

6. **Cleaning:**

a. Thorough cleaning: Once all finishes are in place, clean the entire space thoroughly.

b. Prepare for occupancy: Ensure the space is ready for occupancy or handover.

POSSIBLE ERRORS

Potential errors that may occur at this stage include:

1. **Plastering Errors**: Plastering is crucial for achieving a smooth surface. If not done correctly, it can further highlight imperfections in the blockwork, resulting in an uneven finish. Skilled workmanship and attention to detail are essential to avoid these errors.

2. **Window Installation Errors:** Poorly installed windows can result in gaps between partitions, compromising aesthetics and energy efficiency. Accurate rubber boundary work (such as weather-stripping) around windows is vital to prevent air leaks. Sliding windows must be installed to operate smoothly without derailing.

3. **Door Installation Errors:** Doors serve multiple functions, including access, security, and aesthetics. Common errors include:
   a. irregular and rough fitting of doors
   b. Color variations in wood or plywood materials used for doors.
   c. Misalignment of architraves and frames due to being unplumbed.

4. **Material Quality Issues:** Detecting poor-quality materials before installation is crucial. Issues can arise with plastic tongue and groove (T & G) materials, subpar roofing covers (e.g., plywood), etc. Ensuring the quality of materials helps prevent future problems.

5. **Color Variation Challenges:** Plastic and wooden materials used for finishes may exhibit color variations. Consistent color selection is essential to enhance the visual appeal.

6. **Wood Finish and Polishing:** Proper polishing of wood materials (e.g., cabinetry, wardrobes) is necessary to achieve the required shine and durability, enhancing their appearance.

7. **Flooring Slope:** Achieving a perfectly level floor is essential. Any slope can lead to discomfort and negatively affect furniture placement.

TIPS & OTHER CONSIDERATIONS

1. Inspect each step thoroughly to identify and rectify potential issues early. Address imperfections promptly to ensure a flawless finish.
2. Engage experienced workers with specialized skills for each task. Their expertise ensures high-quality workmanship and minimizes errors.
3. Adhere strictly to quality standards for both materials and workmanship. Prioritize standard materials during procurement. Poor-quality materials yield unsatisfactory results and can cause future problems. Invest in reliable options from the outset to avoid substandard materials.
4. When using plywood, especially of the same veneer type (e.g., Acajou), be mindful of variations in shades. Select plywood with consistent shades to maintain uniformity in your finishes.
5. Consider alternatives and choose the most suitable option based on aesthetics, durability, and cost. Evaluate different materials and methods to find the best fit for your project.

By focusing on these areas, the finishing stage can be completed successfully, resulting in a durable and aesthetically pleasing structure.

CHECKLISTS

Below are useful checklist forms for the site supervisor. Though the checklist is not exhaustive it would serve as a useful guide for the site supervisor. This page can be reproduced and used as a practical guide on-site. Empty spaces provided are for other details the site supervisor would like to checklist.

PRECONSTRUCTION PHASE CHECKLIST

s/n	Process	Started	Completed
1	Review plans with building official.		
2	Make adjustment and modifications		
3	Create/design a site plan.		
4	Obtain a building permit.		
5	Final detailed drawings and specification.		
6	Scheduling of tasks.		
7	Forge contracts		
8	Set date for commencement of project.		

CONSTRUCTION TO FOUNDATION LEVEL PHASE CHECKLIST

s/n	Process	Started	Completed
1	Construction of access road?		
2	Site clearing? Sand fills?		
3	Demolition?		
4	Organise/deliver equipment		
5	Organise/supply materials		
6	Organise/bring labour		
7	Build site office/shed		
8	Dig site well/borehole, latrine		
9	Removal of rubbish/rubble from site		
10	Set out foundation		
11	Dig foundation trench		
12	Blinding & anti-termite?		
13	Formwork/reinforcement?		
14	Lay foundation bed/beam/pad		
15	Foundation wall		
17	Infill blocks with concrete?		
18	Hardcore filling		
19	Screed & DPM?		

s/n	Process	Started	Completed
20	Formwork/reinforcement?		
21	Plumbing & electrical pipes?		
22	Biscuits?		
23	Cast concrete floor		

CONSTRUCTION OF SUPER STRUCTURE PHASE CHECKLIST

s/n	Events/Process	Started	Completed
1	Set out internal and external partitions		
2	Organise/deliver equipment		
3	Organise/supply materials		
4	Organise/bring labour		
5	Block work to lintel level		
6	Plumber piping/(Rough - In)		
7	Lintel, columns, slabs & arcs formwork		
8	Reinforcement lintels, columns, etc.		
9	Electrical piping/ (Rough - In)		
10	Cast columns, slabs, lintels, etc.		
11	Block work to roof level		
12	Plumbing/Piping		
13	Electrical Piping		
14	Windows & doors frames/sub-frames		
15	Roof carcass		
16	Roof cover		

FINISHES, FIXTURES & FITTINGS PHASE CHECKLIST

s/n	Events/Process	Started	Completed
1	Organise/deliver equipment		
2	Organise/supply materials		
3	Organise/bring labour		
4	Rendering/plastering		
5	Floor Covering, Tiling		'
6	Ceiling work		
7	Carpentry (Joinery)		
8	Plumbing fittings		
9	Electrical Fittings		
10	External Work, Landscaping		
11	Cleaning		
12	Final Completion Inspection		
13	Handover		

www.ingramcontent.com/pod-product-compliance
Lightning Source LLC
Chambersburg PA
CBHW040759240526
45474CB00008B/110

*9 7 9 8 3 3 2 6 2 1 9 8 7 *